YOUR KNOWLEDGE HAS VALUE

- We will publish your bachelor's and
  master's thesis, essays and papers

- Your own eBook and book -
  sold worldwide in all relevant shops

- Earn money with each sale

Upload your text at www.GRIN.com
and publish for free

Cosmas Alfred Butele

Sericulture. The Extent of the Conservation and Utilization of Sericigenous Resources Biodiversity in the World

Unveiling Opportunities and Prospects for Developing Countries

GRIN Publishing

Bibliographic information published by the German National Library:

The German National Library lists this publication in the National Bibliography; detailed bibliographic data are available on the Internet at http://dnb.dnb.de .

Imprint:

Copyright © 2012 GRIN Verlag GmbH
Print and binding: Books on Demand GmbH, Norderstedt Germany
ISBN: 978-3-656-37576-0

This book at GRIN:

http://www.grin.com/en/e-book/207282/sericulture-the-extent-of-the-conservation-and-utilization-of-sericigenous

GRIN - Your knowledge has value

Since its foundation in 1998, GRIN has specialized in publishing academic texts by students, college teachers and other academics as e-book and printed book. The website www.grin.com is an ideal platform for presenting term papers, final papers, scientific essays, dissertations and specialist books.

Visit us on the internet:

http://www.grin.com/

http://www.facebook.com/grincom

http://www.twitter.com/grin_com

BUTELE COSMAS ALFRED

COURSE TITLE: SERICULTURE

*THE EXTENT OF THE CONSERVATION AND UTILIZATION OF
SERICIGENOUS RESOURCES BIODIVERSITY IN THE WORLD*

Unveiling Opportunities and Prospects for Developing Countries

**ATLANTIC INTERNATIONAL UNIVERSITY
HONOLULU, HAWAII**

*The extent of the conservation and utilization of sericigenous resources biodiversity:
Unveiling opportunities and prospects for developing countries*

TABLE OF CONTENTS

List of Plates

List of Acronyms

A.D	The year of our Lord, i.e. the number of years since the time of Christ
ADF	African Development Fund
BC	Before Christ
BCE	Before Christian Era
CE	Christian Era
CSR&TI	Central Sericultural Research and Training Institute
CSTRI	Central Silk Technological Research Institute
e.g.	for example
Eds	Editors
et al.	and other people
etc	and other similar things
EU	European Union
ICIPE	International Center for Insect Physiology and Ecology
IFAD	International Funding for Agricultural Development
ISEL	Inuula Silk Estates Limited
JICA	Japan International Cooperation Agency
MAAIF	Ministry of Agriculture, Animal Industry and Fisheries
NGOs	Non-Governmental Organizations
USIL	Uganda Silk Industries Limited
USPA	Uganda Silk Producers Association

Introduction

Sericigenous fauna are animals that are capable of producing a natural and highly valued fibre called silk. According to Prasad *et al.* (2010), the word "Sericigenous" is derived from a Greek word "Seri" meaning Silk. Sericigenous fauna have a wide biodiversity, which can be primarily categorized as insect and non-insect groups. The non-insect group includes the Adriatic mussel (seashell *Pinna nobilis* and related species which produce silk referred to as sea silk) and a Madagascarian spider. The silk of non insect origin is of no importance to textile market as its use is limited in the area of optical instruments. The insect group includes all the silk spinning insects whether foraging on mulberry plant or other non mulberry plants; their silk is the one used in making garments. Therefore, Sericigenous resources include the silk producing animals and the associated plants they feed on. This paper reviews the sericigenous resources biodiversity that have so far been identified and documented and the efforts that have been put in place to conserve and utilize them sustainably. The emphasis here is laid on the Sericigenous insects, the silk moths, and their food plants.

Importance of Sericigenous Resources and Conservation

Sericigenous resources are of great use and benefit to man. Silk is a precious commodity; it is generally used for making expensive and very impressive looking garments (textiles). It is actually called the queen of textiles because of its glittering luster, soft feeling, elegance, durability and tensile properties, unmatched by other textiles whether natural or artificial. Silk can be categorized according to the type of food plant the silkworms feed on. For example, the mulberry silk is produced by the fully domesticated *Bombyx mori* which feeds on mulberry plant while the non mulberry silks are categorized as muga, tropical tasar, temperate/oak tasar and eri silks, depending on the type of the food plant and also the species of the silkworm, and the silkworms are not fully domesticated yet, like the *Bombyx mori*. Therefore the muga, tropical tasar, temperate/oak tasar and eri silks are referred to as wild silks. Sericigenous resources are also used for scientific research and development. For example, the silk moth *Bombyx mori* and its food plant, the mulberry plant, *Morus* species, have been extensively researched, domesticated and developed, over many years, for commercial silk production, a practice called sericulture (Prasad *et al.*, 2010).

Description of the Extent of Sericigenous Resources Biodiversity Conservation and Utilization in the world

There are more than 500 species of wild silkworms in the world, but so far only one fully domesticated silkworm species, *Bombyx mori* L. Only a few of the wild silkworms are of commercial importance to produce cloth (Table 1). The exploitation of commercially important wild silkworms is called wild silk farming. Wild silkworms usually produce a tougher and rougher silk than that from domesticated *Bombyx mori* silkworms. Wild silks are usually harvested after the moths have left the cocoons, cutting the threads in the process so that there is not one long thread as with domesticated silkworms. Wild silks tend to be more difficult to bleach and dye than silk from *Bombyx mori*, but most have naturally attractive colours, particularly the rich golden sheen of the silk produced by the muga silkworm from Assam (India) and is often known as Assam silk. The cocoon shells of wild silk moths are toughened or stabilized either by tanning (cross-linking) or by mineral reinforcements (e.g. calcium oxalate). A new method has been developed, demineralizing, which can remove the mineral reinforcements present in wild silks and enables wet reeling like the commercial silk worm *Bombyx mori*. Although wild silk in general constitutes only about 10% of the total silk output in the world and is being dominated by mulberry silk (Braja, 1999), it can still sustain strong local and international market niches if developed, because of its natural attractive colours, durability, own feel and special appeal (Prasad *et al.*, 2010).

Table 1: Some commercially important wild silk moths and their Taxonomic Families, Silk Description and Geographical Distribution:

Species	Taxonomic Family	Silk Description	Geographical Distribution
1. *Antheraea assamensis* (Helfer, 1837)	Saturniidae	Muga silk: Has a beautiful glossy golden hue which improves with age and washing. It is never bleached or dyed and is stain resistant. It was reserved for the excusive use of royal families in Assam (India) for 600 years.	India (Assam)
2. *Antheraea mylitta* (Drury, 1773)	Saturniidae	Tropical Tasar silk: copperish in colour and coarse in texture	India
3. *Antheraea pernyi* (Guenerin-Meneville, 1855)	Saturniidae	Temperate/Oak Tasar Silk	China and India
4. *Antheraea yamamai* (The "tensan" silk moth) (Guenerin-Meneville, 1861)	*Saturniidae*	Temperate Tasar Silk: It has been cultivated in Japan for more than 1000 years. It produces a naturally white silk but does not dye well, though it is very strong and elastic. It is now very rare and expensive.	Japan
5. *Antheraea roylei* Moore	Saturniidae	Temperate Tasar Silk	India, China
6. *Samia Cynthia* (Drury, 1773) (The Ailanthus Silkmoth): A somewhat domesticated silkworm	Saturniidae	Eri Silk: White silk which resembles wool mixed with cotton, but feels like silk.	India, China and introduced into North America
7. *Samia Cynthia ricini* Donovan: It is a subspecies	Saturniidae	Eri Silk: The silk is extremely durable, but cannot be easily reeled off the cocoon and is	India, China

of *Samia cynthia*		thus spun like cotton or wool.	
8. *Antheraea paphia* Linn.	Saturniidae	Tropical Tasar Silk	India
9. *Anaphe vanata* Butler	Notodontidae	Anaphe Silk	Nigeria
10. *Anaphe Infracta* Wals.	Notodontidae	Anaphe Silk	Nigeria
11. *Anaphe reticulate* Walker	Notodontidae	Anaphe Silk	Uganda
12. *Anaphe panda* Boisduval	Notodontidae	Anaphe Silk	Zaire, Togo, etc.
13. *Epanaphe molonei* Druce	Notodontidae	Anaphe Silk	Nigeria
14. *Epanaphe carteri* Walsingham	Notodontidae	Anaphe Silk	Cameroun
15. *Epanaphe vuilleti* Joan	Notodontidae	Anaphe Silk	Cameroun
16. *Attacus atlas* Linn.	Saturniidae	Fagara Silk	India, China
17. *Attacus cramer* Fldr	Saturniidae	Fagara Silk	India, China
18. *Attacus edwardsi* White	Saturniidae	Fagara Silk	India, China
19. *Attacus doherteyi* Roth	Saturniidae	Fagara Silk	India, China
20. *Attacus standingeri* Roth	Saturniidae	Fagara Silk	India, China
21. *Pachypasa otus* Drury	Lasiocampidae	Coan Silk	Italy, Greece
22. *Pachypasa lineosa* Vill	Lasiocampidae	Coan Silk	Italy, Greece

Sources: Braja, 1999; Mal, 2005; Prasad *et al.*, 2010 and
www.freewebs.com/chinesetussah/antheraeapolyphemus.htm

Table II below shows some wild silkworms that have not yet been commercially exploited but have potential for silk production

Table II: Some important potential wild silk moths and their Taxonomic Families and Geographical Distribution:

Species	Taxonomic Family	Geographical Distribution
1. *Attacus lorquinii* Fldr.	Saturniidae	Indonesia, China, South East Asia
2. *Attacus caesar* Msn.	Saturniidae	Indonesia, China, South East Asia
3. *Attacus inopinatus* J. & L.	Saturniidae	Indonesia, China, South East Asia
4. *Archaeoattacus edwardsii* White	Saturniidae	Indonesia, China, South East Asia

5. *Bombyx mandarina* Moore	Bombycidae	Indonesia, China, South East Asia
6. *Samia watsoni* Ober	Saturniidae	Indonesia, China, South East Asia
7. *Samia carringii* Hutton	Saturniidae	Indonesia, China, South East Asia
8. *Samia walkeri* Fldr.	Saturniidae	Indonesia, China, South East Asia
9. *Samia Yayukae* Pksd & Pglr.	Saturniidae	Indonesia, China, South East Asia
10. *Samia peigleri* Nmn & Nsg	Saturniidae	Indonesia, China, South East Asia
11. *Rhodinia verecunda* Inoue	Saturniidae	Indonesia, China, South East Asia
12. *Fhodinia Jankoswii* Obrtr.	Saturniidae	Indonesia, China, South East Asia
13. *Leopa anthara* Jordan	Saturniidae	Indonesia, China, South East Asia
14. *Leopa miranda* Moore	Saturniidae	Indonesia, China, South East Asia
15. *Leopa megacore* Mell	Saturniidae	Indonesia, China, South East Asia
16. *Cricula haytiae* Pkstd & Shdjn	Saturniidae	Indonesia, China, South East Asia
17. *Caligula Japanica* Shiraki	Saturniidae	Indonesia, China, South East Asia
18. *Caligula Jonasii* Sonan	Saturniidae	Indonesia, China, South East Asia
19. *Caligula boisduvalii* Evsmn	Saturniidae	Indonesia, China, South East Asia
20. *Caligula thibeta* Okano	Saturniidae	Indonesia, China, South East Asia
21. *Caligula zuleika* Hope	Saturniidae	Indonesia, China, South East Asia
22. *Saturnia pyretorum* Watson	Saturniidae	Indonesia, China, South East Asia
23. *Saturnia boisduvalii* Eversman	Saturniidae	Indonesia, China, South East Asia
24. *Actias selene* Fldr.	Saturniidae	Indonesia, China, South East Asia
25. *Actias heterogyna* Kishida	Saturniidae	Indonesia, China, South East Asia
26. *Actias neidhofesi* Ong. & Ya.	Saturniidae	Indonesia, China, South East Asia
27. *Actias maenas* Dbld.	Saturniidae	Indonesia, China, South East Asia
28. *Actias groenendaeli* Roepke	Saturniidae	Indonesia, China, South East Asia
29. *Actias dubernardi* Ober.	Saturniidae	Indonesia, China, South East Asia

30. Actias rhodopneuma Rober	Saturniidae	Indonesia, China, South East Asia
31. Antheraea alleni Holloway	Saturniidae	Indonesia, China, South East Asia
32. Antheraea formosana Sonan	Saturniidae	Indonesia, China, South East Asia
33. Salassa lola Westwood	Saturniidae	Indonesia, China, South East Asia
34. Salassa megastica Swinhoe	Saturniidae	Indonesia, China, South East Asia
35. Rhodinia fugax Butler	Saturniidae	Indonesia, China, South East Asia
36. Rhodinia Jankowskii Ober.	Saturniidae	Indonesia, China, South East Asia
37. Antheraea polyphemus Linn.	Saturniidae	America (North America): *Antheraea polyphemus* has the most potential of any North American wild silkworms)
38. Callosamia promethia Drury	Saturniidae	America (North America)
39. Gonometa postica Walker	Lasiocampidae	Africa (Kalahari region)
40. Gonometa rufobrunnae Arvls	Lasiocampidae	Africa (Uganda)
41. Hyalophora cecropia Linn. (The quality of the silk depends on food source).	Saturniidae	America (North America)
42. Eutachyptera psidii Salle	Lasiocampidae	America
43. Eucheria socialis Westwood (The only butterfly silk)	Pieridae	America
44. Malacosoma Incurvum Aztecum	Lasiocampidae	America
45. Malacosoma americanum Fabr.	Lasiocampidae	America
46. Antheraea montezuma Salle	Saturniidae	America
47. Antheraea godmani Druce	Saturniidae	America
48. Hyalophora euryalus Boisdv	Saturniidae	America
49. Hyalophora Columbia Strecker	Saturniidae	America
50. Actias luna Linn.	Saturniidae	America
51. Saturnia walterorum H. & J.	Saturniidae	America
52. Saturnia pyri D. & S.	Saturniidae	America
53. Borocera cajani Vinson	Lasiocampidae	Africa
54. Argema Sp.	Saturniidae	Africa (Uganda)
55. Antistathmomoptera Sp.	Saturniidae	Africa
56. Gonimbrasia Sp.	Saturniidae	Africa
57. Atheletes Sp.	Saturniidae	Africa
58. Bunaeopsis Sp.	Saturniidae	Africa (Uganda)
59. Imbrasia Sp.	Saturniidae	Africa (Uganda)
60. Bunaea Sp.	Saturniidae	Africa (Uganda)

Sources: Braja, 1999; Nanny, *et al.*, 2004; and
www.freewebs.com/chinesetussah/antheraeapolyphemus.htm

Table 3: Biodiversity of *Antheraea* species in the world:

Name of species	Name of species
1. *Antheraea assamensis* (Helfer, 1837)	21. *Antheraea gephyra* Niep.
2. *Antheraea mylitta* (Drury, 1773)	22. *Antheraea raphrayi* Bouv.
3. *Antheraea pernyi* (Guenerin-Meneville, 1855)	23. *Antheraea imperator* Wts.
4. *Antheraea yamamai* (Guenerin-Meneville, 1861)	24. *Antheraea eucalypti* Scott
5. *Antheraea roylei* Moore	25. *Antheraea billitonensis* Mr.
6. *Antheraea paphia* Linn.	26. *Antheraea prelarissa* Bouv.
7. *Antheraea alleni* Holloway	27. *Antheraea knyvetti* Hmps.
8. *Antheraea formosana* Sonan	28. *Antheraea sivalika* Mr.
9. *Antheraea polyphemus* Linn.	29. *Antheraea crompta* R. & J.
10. *Antheraea montezuma* Salle	30. *Antheraea semperi* Fldr.
11. *Antheraea godmani* Druce	31. *Antheraea ridlei* Mr.
12. *Antheraea frithii* Mr.	32. *Antheraea surakarta* Mr.
13. *Antheraea helferi* Mr.	33. *Antheraea pratti* Bouv.
14. *Antheraea andamana* Mr.	34. *Antheraea harti* Mr.
15. *Antheraea janna* Stoll	35. *Antheraea pasteuri* Bouv.
16. *Antheraea Larissa* Ww.	36. *Antheraea cordifolia* Weym.
17. *Antheraea pristine* Wkr.	37. *Antheraea brunnea* Eecke.
18. *Antheraea delegate* Swh	38. *Antheraea larissoides* Bouv.
19. *Antheraea mylittoides* Bouv.	39. *Antheraea sciron* Ww.
20. *Antheraea rumphi* Fldr.	40. *Antheraea fiekei* Weym.

Sources: Braja, 1999; Mal, 2005 and www.freewebs.com/chinesetussah/antheraeapolyphemus.htm

The domesticated silkworm *Bombyx mori* is fed exclusively on mulberry plant, *Morus* species leaves (Plate VI) cultivated in plantations. *Bombyx mandarina* also feeds on mulberry. There are several species and varieties of *Morus* e.g. *M. nigra, M. multicaulis, M. australis, M. alba, M. alba var. macrophypllu,* and *M. bombycis* (Kartasubrata, 2005). The eri silk worm, *Samia ricini,* from India feeds mainly on the leaves of the castor plant, *Ailanthus altissima;* sometimes *Kesseru,* Tapioca/Cassava, Papaya, *Payam* and *Barkesseru* are used. It is the only nearly domesticated silkworm next to *Bombyx mori. Philosomia cynthia pryeri* from China feeds on *Ailanthus altissima, Picrassma quassiodes,* etc. *Antheraea pernyi* from China feeds on *Quercus acutissima,* oak, *Quercus dentate, Quercus serrata,* etc. *Anaphe reticulate* from Uganda feeds on bridelia trees (*Ficus* species). The primary food plants of *Antheraea paphia* and *A. proylei* include *Shorea robusta, Anogeissus latifolia, Terminalia tomentosa, T. arjuna, Lagerstroemia parviflora* and *Madhuca indica,* but there is a wide biodiversity of food plants of *Antheraea paphia* (Table IV). The caterpillars of *A. assamensis,* quite similar to *A. paphia* and *A. proylei* and producing Muga Silk thrive well on *Som* (*Machilus bombycina*), *Magnolia hypoleuca* and *Soalu* (*Litsea polyantha*). *Antheraea yamamai* from Japan feeds on *Quercus acutissima,* oak, *Quercus dentanta. Dictioproca japonica* from Japan feeds on *Jugulans mandshurica, Castanea crenata, Cinnamonum camphora,*

Rhus verniciflus, etc. Anaphe moloneyi, Anaphe infracta and *Anaphe venata* all from Africa also all feed on *Ficus* species (Braja, 1999; Mal, 2005; and Mugyenyi, 2006). Having a variety of silkworm species and their food plants is a good genetic resource for stability and sustainability in the rapidly changing environment.

Table IV: Biodiversity of food plants of the tropical tasar silkmoth, *Antheraea paphia:*

Food plant species	Food plant species
1. *Terminalia tomentosa*	*27 Lagerstroemia indica*
2. *Terminalia arjuna*	*28 Lagerstroemia speciosa*
3. *Terminalia catapa*	*29 Zijiphus jujube*
4. *Terminalia belerica*	*30 Ziziphus mauritiana*
5. *Terminalia glabra*	*31 Ziziphus xylopyra*
6. *Terminalia foetidissima*	*32 Ziziphus rugosa*
7. *Terminalia manti*	*33 Ficus religiosa*
8. *Terminalia myriocarpa*	*34 Ficus bseila*
9. *Terminalia procera*	*35 Ficus retusa*
10. *Terminalia mucronata*	*36 Ficus benjamina*
11. *Terminalia chebula*	*37 Bauhinia variegate*
12. *Terminalia paniculata*	*38 Bamby malbaricum*
13. *Terminalia pyrifolia*	*39 Bamby neptaphylum*
14. *Terminalia muellerian*	*40 Hardwickia binata*
15. *Terminalia utrina*	*41 Melostoma malbaricum*
16. *Terminalia pallida*	*42 Careya arborea*
17. *Terminalia bialata*	*43 Casuarina equisetifolia*
18. *Tectona grandis*	*44 Carissa caranda*
19. *Madhuca indica*	*45 Rhizophora caleolaris*
20. *Anogeissus latifolia*	*46 Pentaptera tomentosa*
21. *Rosa indica gigantean*	*47 Pentaptera glabra*
22. *Shorea robusta*	*48 Carsia lanceolata*
23. *Shorea talura*	*49 Dodonea viscose*
24. *Syzygium cumuni*	*50 Webera corymbosa*
25. *Syzygium sphoerica*	*51 Cipadessa fructose*
26. *Lagerstroemia parviflora*	*52 Cantium didynum*

Sources: Braja, 1999; Mugyenyi, 2006; and www.freewebs.com/chinesetussah/antheraeapolyphemus.htm

General Analysis of the Extent of Sericigenous Resources Biodiversity Conservation and Utilization in the world

There is eminently a wide biodiversity of Sericigenous resources (Tables 1, II, 3 and IV; Plates 1, II, 3, IV and 5). Only a few of them (Table 1 and Plates 1, II, 3 and IV) such as the *Bombyx mori* silkworm and the associated food (mulberry) plant, *Morus* species, have been singled out for continuous conservation and utilization over the years (Tables 5, VI and 7). This has been driven mainly by the increasing desire and demand for their silk and economic gain (Plates VIII (a)-(d) and 9) (Dandin, 2005). To keep in the pace, there has also been a continuous search for the best qualities in mulberry leaf yield, and the cocoon and the silk produced. In the process, the original gene in the original stock has been altered to create a number of varieties with different qualities. These few

Sericigenous resources already proved to be most commercially viable have blindfolded the world from looking at the others, with over 500 silk moth species being still in the wild (Braja, 1999). Although the manipulation of these two resources (*Bombyx mori* and *Morus* species) has generated several man-made varieties, the overdependence and overexploitation of them, as is monoculture in nature (Plate 7), is dangerous and could lead to loss of the desired characters "true to the species" and the eventual disappearance from the scene as has already been observed with the ancestral *Bombyx mandarina*, hence the loss of Sericigenous resources biodiversity. The ever-increasing human population and the resultant pressure on natural resources generally, coupled with looming climatic change even make the situation more precarious. Particular *Bombyx mori* races (Mal, 2005) and the *Morus* species varieties or cultivars (Ravindran and Rajanna, 2005) are being promoted and utilized in many parts of the world for silk production. The research and development work that developed them has, however, been confined in the laboratories of few countries mainly in Asia. That means their adaptation to and survival in new environments in the new twist of global warming will be less than the indigenous or local species in those areas. It is now important to begin drawing attention to conserving and developing the indigenous Sericigenous resources alongside the exotic ones for cross-breeding and complimenting, while maintaining a reserve of the original stock. Wild silkworm rearing approach would even be the most appropriate in less advantaged rural communities where level of education and technology adoption is still low. Such a practice, acting as a strategy at the same time, would conserve Sericigenous flora and fauna, forests and the general biodiversity.

Actualization: Practical case studies of *Bombyx mori* Conservation and Utilization in China, India and Uganda

Although wild silk farming started in China about 4,000 years ago, the present day mulberry silk from *Bombyx mori* was discovered in a Chinese legend 2,600-2,700 BC. Following is an account of the efforts to conserve and utilize *Bombyx mori* since then, in China, India and Uganda.

Table 5: Chronology of the conservation and utilization of *Bombyx mori* in China:

Year	Event
5,000-3,000 BCE	• Silk cultivation started in China using wild silk worms, during Yangshao period.
2,600-2,700 BC	• Mulberry silk from wild *Bombyx mori* (Linn.) was discovered during the reign of the great emperor of China, Huang Ti, as per legend. *Bombyx mori* evolved from its wild ancestor *Bombyx mandarina* Moore
2,700 BC	• Modern sericulture started with the domestication and improvement of the *Bombyx mori* silkworm.
2,200 BC	• First official documentation on silk by Chinese king "Chou King".
2,200 BC-195 A.D	• Sericulture spread through China • Silk became a very precious commodity

	• highly sought by other countries to make very expensive and impressive looking garments • Chinese kept the secret from other countries for a long time • China was the sole producer and exporter of silk through the so called silk road
195 A.D	• Sericulture eventually reached Japan through Korea • Chinese lost the monopoly on silk production
300 A.D	• Silkworm cultivation was established in India. Legend has it that the egg of the insect and the seed of the mulberry tree were carried to India concealed in the headdress of a Chinese Princess.
522 A.D	• Secrets of sericulture reached to Roman Empire in Europe in the reign of Emperor Justinian through smuggling of the silkworm eggs from China by Persian monks • Later, sericulture was introduced to other European, Mediterranean and Asiatic countries

Sources: Dandin, 2005; Mal, 2005; MAAIF, 2007; and, Prasad *et al.*, 2010

Sericulture has also been attempted in the United States, but these endeavours have been sporadic and largely unsuccessful except in Brazil (Raje, 1999 and Ron, 2011). In Africa, mulberry sericulture has recently been started in Uganda, Kenya, Tanzania, Ethiopia, Ghana, Botswana, Cote d'Ivore and Rwanda. Not much has been achieved in sericulture in the continent but with the advantages of suitable climate, large amounts of arable land and labour at reasonable cost and the growing demand for affordable silk products, Africa could become a major supplier of silk in the world in future (MAAIF, 2007).

Table VI: Chronology of the conservation and utilization of *Bombyx mori* in India

Year	Event
1670	• Modern era silk trade was started in by East India Company
1900s	• British Government took measures to expand sericulture in India • Tippu Sulthan established Mysore Silk Industry • James Anderson introduced sericulture in Madras Presidency
1911	• Department of sericulture started in Karnataka
1919	• Department of Sericulture started in Madras
1943	• A small Research Institute was established in by British Government in Berhampore (West Bengal)

1945	• Government of India created a Silk Development Directorate
1947	• All India Silk Conference was organized
1948	• Central Silk Board of India was established
1960	• Central Research Coordination Committee was constituted
1961	• Central Sericultural Research and Training Institute (CSR&TI), Mysore was established
1964	• CSR&TI, Ranchi (Bihar) was established
1969	• Berhampore Research Institute was converted into CSR&TI • CSB established a number of regional stations and extension centres in different states • State governments strengthened official networks for sericultural development in the country
1963-70	• Package of practices for mulberry cultivation were developed by CSR&TI
1970s	• Four exogenous bivoltine races of silkworms (NB7, KA) from China (oval) and (NB4D2 and NB18) from Japan (dumbbell) were introduced.
1982	• Central Silk Technological Research Institute (CSTRI) was established at Bangalore
1990s	• High yielding mulberry variety (VI) was evolved • 3 new indigenous bivoltine races (CSR2, CSR4 and CSR5) evolved and 2 hybrids (CSR2 X CSR4 and CSR2 X CSR5) and reciprocals were prepared

Source: Prasad *et al.*, 2010

Table 7: Chronology of the conservation and utilization of *Bombyx mori* in Uganda

Year	Event
1920s	• Research trials on mulberry and *Bombyx mori* started, but failed due to lack of silkworm eggs, storage facilities and trained indigenous manpower
1970s	• Attempts were again made by Japanese Government to introduce sericulture with mulberry plants, seeds, grafting twigs and cuttings from Japan to improve local mulberry varieties, and proved successful and mulberry plantations were established at Kawanda Agricultural Research Institute. However, it was only limited to mulberry agronomic practices without silkworm rearing trials, and it stopped in 1975 due to lack of silkworms eggs and trained personnel and political instability
1985	• Indian Government and Swiss Development Corporation provided a training opportunity to one Ugandan at the International Center for

	Training and Research in Tropical Sericulture, Mysore. Later, other Ugandans also trained at this institute. • Silkworm rearing activities were revived at Kawanda Agricultural Research Institute • Sericulture unit established at Kawanda under the Ministry of Agriculture, Animal Industry and Fisheries (MAAIF)
1986-1989	• Successful rearing trials were conducted using silkworm eggs imported from India • Evaluation and multiplication of mulberry varieties carried out
1989	• A team of Indian Sericulture experts identified Bushenyi District as the most promising area for on-farm activities
1990	• Two private companies; Uganda Silk Industries Limited (USIL) in the Central and Western Uganda and Inuula Silk Estates Limited (ISEL) in the Eastern Uganda initiated commercial cocoon production mainly for export.
1992	• Commercial cocoon production started. • But USIL and ISEL failed due to technical, environmental and financial reasons
1993	• EU sponsored a project to assist private companies which are members of Uganda Silk Producers Association (USPA) and promote commercial sericulture
1994-1999	• There were already 388 cocoon producing farm households • 22 District Sericulture Centres established • Dry silk cocoons of grade 2A were exported (70%) to Japan • International market demand shifted for the semi processed product, the silk yarn instead of the silk cocoons • Silk industry in Uganda destined to crush again
1999-2002	• International Centre for Insect Physiology and Ecology (ICIPE) started a trial on cocoon processing and promoted raw silk production with funding from International Funding for Agricultural Development (IFAD).
2002	• Bushenyi Silk Factory was established under ICIPE/IFAD funding
2003	• There were 105 active cocoon producers • Silk reeling machine established at Kawanda under African Development Foundation (ADF) funding
2004	• There were 133 active cocoon producers
2005-2012	• Japan International Co-operation Agency (JICA)

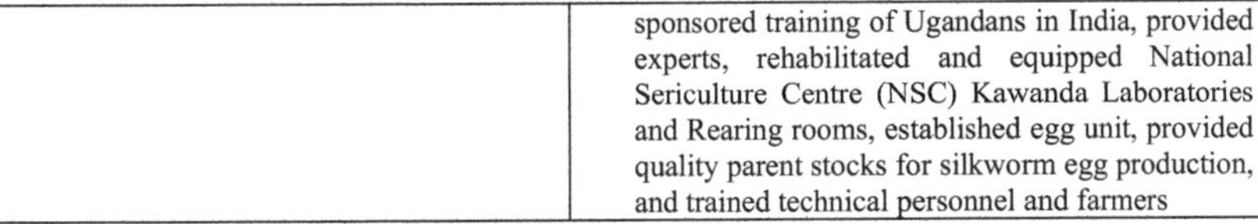

	sponsored training of Ugandans in India, provided experts, rehabilitated and equipped National Sericulture Centre (NSC) Kawanda Laboratories and Rearing rooms, established egg unit, provided quality parent stocks for silkworm egg production, and trained technical personnel and farmers

Sources: Mugyenyi, 2006; MAAIF, 2007; and MAAIF, 2008

Actualization: Practical Case Studies of Wild Silkworm Conservation and Utilization in the world

Wild silk farming has been developed to a great extent in India and wild silks are often referred to in India as "Vanya" silks. The term "Vanya" is of Sanskrit origin, meaning untamed, wild, or forest-based. Muga, Tassar and Eri silkworms are not fully tamed and the world lovingly calls the silks they produce as "wild silks". India produces all the 4 commercially exploited wild silk varieties, namely; Tropical Tasar, Temperate/Oak Tasar, Eri and Muga and is the only country in the world producing all the 5 silk varieties, including Mulberry silk, because of the diverse climatic zones and abundant natural resources in the country (Prasad *et al.*, 2010). The tropical tasar silkmoth, *A. paphia*, is exclusively cultivated in India and the estimated annual production of tasar silk is 130 tonnes while the production of other types of wild silk exceeds 10,000 tonnes. Wild silk farming still has a long future because of its attachment to the tribal communities, socio-economic benefits and being forest based and enhancing environmental conservation. Historically, wild silk threads have been found and identified from two Indus River sites, Harappa and Chanhu-daro, dating to c. 2450-2000 BCE. This is roughly the same period as the earliest evidence of silk use in China, which is generally thought to have had the oldest silk industry in the world. The specimens of threads from Harappa appear on Scanning electron microscope analysis to be from two different species of silk moth, *Antheraea mylitta* (=*A. paphia*) and *A. assamensis*, while the silk from Chanhu-daro may be from a *Philosamia (=Samia)* species, Eri Silk, and this silk appears to have been reeled. Wild silks were in use in China from early times. Moreover, the Chinese were aware of their use in Roman Empire and apparently imported goods made from them by the time of the Later Han Dynasty in the 1st to 3rd centuries CE. There are significant indications in the literature that wild silks were in use in Persia and in Greece by the late 5th century BCE, apparently referred to as "Amorgina" or "Amorgian garments" in Greece. Pliny the Elder, in the 1st century CE, obviously had some knowledge of how wild silkworms' cocoons were produced and utilized on the island of Kos, even though his account included some fanciful ideas. In Uganda, indigenous wild Anaphe silk cocoons were being collected from indigenous wild bridelia trees between 1910 and 1945 and exported (9 tonnes of cocoon per year) to Europe (Mugyenyi, 2006).

Discussions: Advantages and Disadvantages of Sericigenous Resources Biodiversity Conservation and Utilization

Sericigenous fauna and flora play a major role in an ecosystem by enriching the biodiversity and balancing the ecosystem. Therefore, their conservation is vital in environmental protection besides using them for study or research purposes and silk production for income earning or as human or livestock feed. A good number of the local community members can be employed in conservation and cultivation activities of Sericigenous resources. Despite their benefits, the extent of conservation of Sericigenous resources is still low world over. Only one (i.e. *Bombyx mori*) out of over 500 known wild silk moths in the world has so far been domesticated and conserved. It is feared that some of the potentially important indigenous wild silk moths may get extinct with time and change if strategic conservation measures are not employed in time. Although sericigenous resources are widespread across the world, not all communities have a strong association with and attachment to especially the silk worms like the Chinese, Japanese and Indians who have had a long time attachment to the caterpillars (Tables 1 and VI). In fact, in many communities in Africa, the sight of caterpillars scares. This means that traditional or cultural attitude is paramount and should be considered when introducing conservation efforts into new areas. Despite the wide geographical distribution of Sericigenous resources, the world appears to be depending solely on those that have already been developed elsewhere. First of all, this may leave much of this resource at our disposal in the rural communities untapped. Secondly, the indigenous species and their characters will not be used to stabilize and standardize exogenous species introduced into a new area as is the case in India (Table VI). This is therefore a challenge for Governments, universities, research institutions, NGOs and donors with interest in poverty reduction and sustainable development especially so in developing countries like Uganda.

General Recommendations

The highly developed *Bombyx mori* silkworm races should be evolved to be able to adapt to the looming global warming and tropical conditions as it can no longer survive in the wild even where it originated from.

The search for food sources for the *Bombyx mori* silkworm other than the *Morus* species should continue as moriculture (mulberry cultivation) is predominantly monoculture in practice with its ecological setbacks and as the information about alternative sources of food for *Bombyx mori* is scanty.

Already established commercially important wild silk moths should also be developed to complement *Bombyx mori* in case of an unforeseen calamity like disease outbreak on *Bombyx mori* as it is feared to be the only surviving member of its family Bombycidae.

The search for other non-insect silk fauna should continue as information about them is scanty.

The search for indigenous Sericigenous resources should continue in the countries like Uganda that have not been pioneering in silk industry, to be able to use the genes to stabilize and standardize the exogenous species, and to add on to the Sericigenous resources base and the existing body of knowledge in sericulture

Conclusion

It is only *Bombyx mori* silk moth out of over 500 known wild silk moths in the world that has so far been fully domesticated and utilized continuously for silk production over the years. This species is likely to lose its "true to the species" parental characters with time.

In some communities in Africa, the sight of caterpillars scares people. A traditional, cultural or societal attitude such as this will affect introduction of Sericigenous resources conservation projects in new areas.

Wild silk farming approach gives both the ecological conservation and socio economic benefits to the community. It can easily be adopted by the rural communities along forests.

Bombyx mori is monophagous on mulberry and information on alternative sources of food is scanty. This can affect *Bombyx mori* conservation and utilization efforts adversely in case of disaster strike on mulberry.

Only 6 indigenous silk moth species from Uganda have been captured in this review. This is a reflection of lack of deliberate research and information published on particularly the indigenous silk moths in Uganda. This review has therefore highlighted the enormous opportunity for further research in future on the indigenous Sericigenous resources in Uganda.

Bibliography

- BRAJA, KISHORE NAYAK. "Wild Silkmoth Farming for Income Generation and its impact on Biodiversity". <u>The Conservation and Utilization of Commercial Insects.</u> Proceedings of the First International Workshop on the conservation and utilization of commercial insects held in Nairobi, Kenya, 18-21, August 1997. Eds Raina, S.K.; Kioko, E.N.; and Mwanycky, S.W. ICIPE Science Press. 1999: p125-142.
- DANDIN, S.B. "About the Book…" <u>Silkworm Breeding & Genetics.</u> H. Basker, I.A.S. Bangalore, India, 2005.
- http://en.wikipedia.org/wiki/sericulture#mw-head
- http://www.en.wikipedia.org/wiki/File:Muga_Silkworm.JPG
- http://www.en.wikipedia.org/wiki/File:Silkworm_&_cocoon.jpg
- http://www.en.wikipedia.org/wiki/Wild_silk#History
- http://www.en.wikipedia.org/wiki/Wild_silk#List_of_some_wild_silk_moths_and_their_silk
- http://www.en.wikipedia.org/wiki/Wild_silk#Wild_silk_industry_in_India

- KARTASUBRATA, J. "Uses of *Morus* L. in Sericulture". PROSEA Newsletter. Number 34. April 2005. www.proseanet.org/prosea/_newsletters/newsletter_34_april2005.htm
- MAAIF. A National Strategy for Sericulture Development 2006-2016. 2007.
- MAAIF. National Policy for Sericulture Development. 2008.
- MAL, REDDY N. "Silkworm races". Silkworm Breeding & Genetics. H. Basker, I.A.S. Bangalore. India. 2005: p21-38.
- MUGYENYI, G. Sericulture Development in Uganda: Policy Issues and Strategies. A paper presented at Workshop on JICA's Co-operation in Agricultural Sector on 29th August 2006, Hotel Africana, Kampala Uganda. 2006.
- NANNY, CARDER; TINDIMUBONA, LAURA; and TWEIGYE, CHARLES K. Butterflies of Uganda. Second edition. Uganda Society. Kampala. Uganda. 2004.
- PRASAD, G.V.; MOGILI, T.; RAGHUPATHI, M.; SATHYANARAYANA, RAJU; and QADRI, S.M.H. (eds). Sericulture Field Guide. Vjayavani Printers. Chowdepalle. India. 2010.
- RAJE, URS S. "Strategy for Development of Grainage for Silkmoth Eggs in African Conditions". The Conservation and Utilization of Commercial Insects. Proceedings of the First International Workshop on the conservation and utilization of commercial insects held in Nairobi, Kenya, 18-21, August 1997. Eds Raina, S.K.; Kioko, E.N.; and Mwanycky, S.W. ICIPE Science Press. 1999: p43-48.
- RAVINDRAN, S. and RAJANNA, L. "Mulberry Production Management". Mulberry Cultivation & Physiology. H. Basker, I.A.S. Bangalore. India. 2005: p1-54.
- RON, CHERRY. "History of Sericulture". Insects.org. 2011. http://www.insets.org/ced1/seric.html
- SURESH, KUMAR N. "Breeding Techniques". Silkworm Breeding & Genetics. H. Basker, I.A.S. Bangalore. India. 2005: p39-82.
- www.freewebs.com/chinesetussah/antheraeapolyphemus.htm

Plates

Plate1: *Bombyx mori* silkworms feeding on mulberry leaves
(Photo by Cosmas Alfred Butele).

Plate II: *Bombyx mori* silkworm and the cocoon
(Source: http://www.en.wikipedia.org/wiki/File:Silkworm_&_cocoon.jpg)

Plate 3: Muga silkworms on a Som tree
(Source: http://www.en.wikipedia.org/wiki/File:Muga_Silkworm.JPG)

Plate IV: *Anaphe reticulate* silkworms on bridelia tree leaves
(Photo by Christine Asaba)

Plate 5 (a): Adult *Antheraea polyphemus*
(Source: *www.freewebs.com/chinesetussah/antheraeapolyphemus.htm*)

Plate 5(b): Adult female and male *A. polyphemus* mating
(Source: *www.freewebs.com/chinesetussah/antheraeapolyphemus.htm*)

Plate 5 (c): Eggs of *A. polyphemus*
(Source: *www.freewebs.com/chinesetussah/antheraeapolyphemus.htm*)

Plate 5(d): First instar larva of *A. polyphemus*
(Source: *www.freewebs.com/chinesetussah/antheraeapolyphemus.htm*)

Plate 5(e): Second instar larva of *A. polyphemus*
(Source: *www.freewebs.com/chinesetussah/antheraeapolyphemus.htm*)

Plate 5(f): Third instar larva of *A. polyphemus*
(Source: *www.freewebs.com/chinesetussah/antheraeapolyphemus.htm*)

Plate 5(g): Fourth instar larva of *A. polyphemus*
(Source: *www.freewebs.com/chinesetussah/antheraeapolyphemus.htm*)

Plate 5(h): Fifth instar larva of *A. polyphemus*
(Source: *www.freewebs.com/chinesetussah/antheraeapolyphemus.htm*)

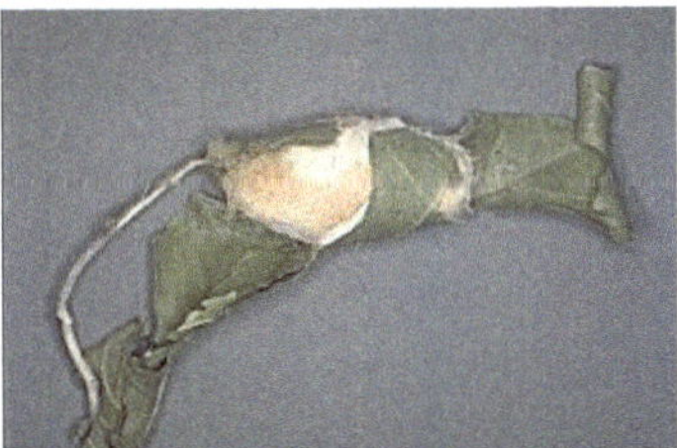

Plate 5 (i): Cocoon of *A. polyphemus*
(Source: *www.freewebs.com/chinesetussah/antheraeapolyphemus.htm*)

Plate VI: Leaves of *Morus* species. (Source: Kartasubrata, 2005)

Plate 7: Mulberry plantation established for *Bombyx mori* silkworm rearing in India (Photo by Cosmas Alfred Butele)

Plate VIII (a): *Bombyx mori* silkworm rearing in India
(Photo by Cosmas Alfred Butele)

Plate VIII (b): *Bombyx mori* cocoons in the market in India
(Photo by Cosmas Alfred Butele)

Plate VIII (c): Decorations from silk cocoons.
(Photo by Cosmas Alfred Butele)

Plate VIII (d): Silk yarn ready for weaving into cloth
(Photo by Christine Asaba)

Plate 9: Silk garments from Uganda (Photo by Christine Asaba)